Ebtisam Elghblawi

Inflammoscopy and Sethoscopic of inflammatory skin lesions

Ebtisam Elghblawi

Inflammoscopy and Sethoscopic of inflammatory skin lesions

LAP LAMBERT Academic Publishing

Cover image: www.ingimage.com

Publisher:
LAP LAMBERT Academic Publishing
is a trademark of
Dodo Books Indian Ocean Ltd. and OmniScriptum S.R.L publishing group

120 High Road, East Finchley, London, N2 9ED, United Kingdom
Str. Armeneasca 28/1, office 1, Chisinau MD-2012, Republic of Moldova, Europe
Managing Directors: Ieva Konstantinova, Victoria Ursu
info@omniscriptum.com

Printed at: see last page
ISBN: 978-620-9-54699-0

Inflammoscopy and Sethoscopic of inflammatory skin lesions

Corresponding author: Dr Ebtisam Elghblawi, MBBCh, MScRes, ADD, DRH, Dip dermatoscope. ebtisamya@yahoo.com

The term Inflammoscopy merely implies trichoscopic distinctive features of inflammatory skin conditions; the commonest ones along with some rare ones, to demonstrate the usefulness, and to ease our understanding and help conclude, if possible, in aiding and supporting the diagnosis when it is equivocal. Previously, dermoscopy, an optical device that is used to detect skin malignancy, but promptly, it is also used to analyse inflammatory skin disorders as a supportive tool in various non-neoplastic dermatoses such as inflammatory, infiltrative, and infectious diseases, however; it doesn't follow a standardised approach. It is non-invasive, easy to apply, and allows visualization of the epidermis, dermo-epidermal junction, and papillary dermis, in vivo magnification of the skin which cannot be seen with the naked eye.

Dermoscopy employs a ×10 to ×100 microscopic visualisation with a light source to magnify the structures under the skin's surface. There are three kinds of dermoscopy: namely, conventional nonpolarized dermoscopy, nonpolarized contact dermoscopy, and polarized contact dermoscopy. Many dermoscopic structures have a high degree of correspondence with pathognomonic histopathologic features.[64]

It's considered the dermatologist's stethoscope as it's an economic tool, yet data on the skill of colour is still limited and is a challenge. Histology is the cutting edge to learn about the underlying issues, but that needs a biopsy. Thus, applying trichoscopy features with the clinical findings can be sufficient to establish the right diagnosis without the need for a biopsy. Nonetheless, dermoscopy requires special training and experience.

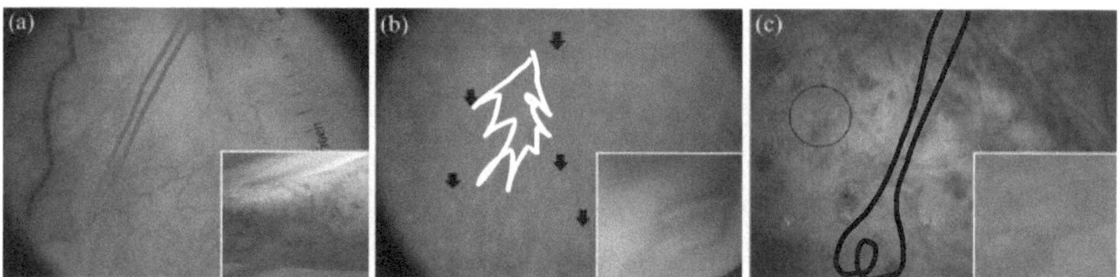

Commonest structures: vessels, scales, hair follicles.

Introduction

The most common structures seen in any inflammatory skin disease are mainly blood vessels (a), scales or crusts (b), and hair follicle changes (c). Thus, the International Dermoscopy Society came up with a consensus for dermatoscopic parameters to be applied when it's applied on the skin in general dermatology usage; namely,

1. blood vessels morphology and distribution,
2. scales, colour and distribution,
3. follicular findings and their disturbances,
4. other specific clues, like in lichen planus (Wickham striae) and porokeratosis (peripheral collarette, which coincides with coronoid lamella histologically).

So, of note, in any inflammatory skin condition, the main thing is to consider the vascular structure along the scales. Applying a noncontact polarising dermatoscope is recommended, in some cases along an interface fluid such as oil or gel to enhance visualization over scales.

Common inflammatory scalp diseases, such as psoriasis, seborrheic dermatitis, lichen planopilaris, discoid lupus erythematosus, contact dermatitis, syphilis, or pemphigus may share similar clinical features. Thus, trichoscopy may offer a quick, non-invasive, cost-effective, and in-office diagnostic method of significant value in clinical practice, with a microstructure view.

The **vessel morphology** can be:

Dotted, as rounded vessels of any size, and is seen in most skin inflammatory conditions, including dermatitis, lichen planus, pityriasis rosea, and porokeratosis. However, **linear** vessels without branching or **curving** can be seen in sun-damaged skin, steroid-treated areas for a long time, and rosacea, which displays a polygonal vessel morphology. On the other hand, any **linear branching** vessels are distinctively associated with basal cell carcinoma, granuloma (tuberculosis and sarcoidosis), and advanced lupus. Besides, **linear curved** vessels (comma vessels) are linked to dermal nevi, lichen planus, and mycosis fungoides[1].

Vessels can be distributed in a distinctive **regular** manner all over the lesion surface, which is typified in psoriasis, or seen at the peripheral as in lichen planus. It can be **patchy** that is randomly arranged without a specific pattern is seen in dermatitis and pityriasis rosea or network (plexus) like in psoriasis (dotted) and rosacea (linear)[1-4].

Scales play a role as well as colour and their distribution can aid in diagnosis as well. **Whitish** is seen in psoriasis mainly and lichen planus, while **yellowish** is linked to serum extravasation and keratin which is seen in all dermatitis coinciding with spongiosis histologically. Also, scale distribution can aid as well, like if diffusely covering the lesion surface which is visualised in several hyperkeratotic dermatoses, while central scales are seen mostly in psoriasis, though it's not specific. Peripheral scales can be appreciated in pityriasis rosea and tinea, while patchy scales can be seen in several disease[1-4].

Hair follicle distribution and its disturbance can aid in diagnosis as well. If it has follicular plugs, which are keratin plus to the hair follicle, this can denote mostly discoid lupus. Additionally, it can show peri-follicular halo or pigmentation, like in alopecia, and repigmentation in vitiligo.

Specific clues can suggest a strong association with specific skin conditions, like the white crossing lines of Wickham striae in lichen planus and the peripheral keratotic rim of porokeratosis.

Psoriasis:

It's a chronic and commonly encountered relapsing skin condition in the clinical setting that is characterised by thick silvery scales mostly in the scalp, and extensor parts of the body, including the flexures and nails. Psoriasis is a multisystem disease.
The trichoscopic features can be: diffuse scaling, simple and twisted red loops, loads of red dots evenly distributed, globules, perifollicular scales, and glomerular vessels in a regular distribution along dry regular white scales observed (bushy pattern). Additionally, hidden hair and signet ring vessels have been reported recently. The uniformly distributed dotted vessels correspond histologically to dilated capillaries in regularly elongated dermal papillae over a dull red background along diffuse scaling, which corresponds histologically to parakeratosis. The vessels are uniform in size and shape, and on scale removal, red dots are noted that correspond to Auspitz sign[5].

Psoriasis, regular distribution[5].

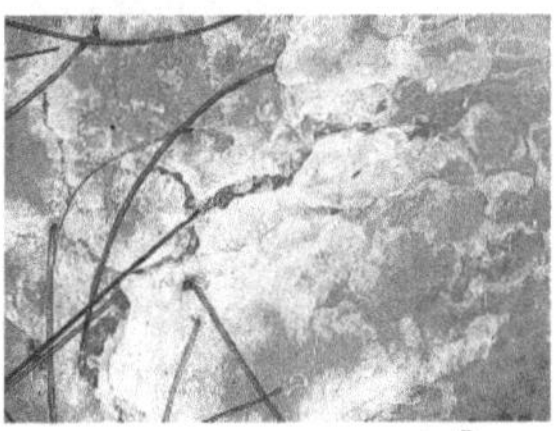

Psoriasis: thick white scales[5].

Eczema/ dermatitis:

It's a commonly encountered relapsing skin condition in the clinical setting that is characterised by itching, and dermatoscopic features are yellow scales/ crusts, which coincide with spongiosis /exocytosis in patchy/ clustered distribution, serocrusts, resulting from hyperkeratosis with some red dots but not linear, and white halo in chronic phases (lichenification)[1].

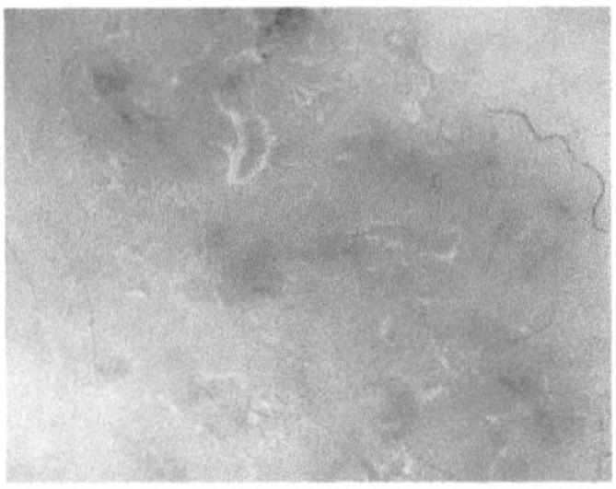

Dermatitis: dotted vessels in clusters with yellow scales[1].

Discoid eczema, nummular eczema:

The predominate dermaoscopy feature noted were dotted vessels in irregular distributions along white scales over yellowish scales with yellow structureless areas and "sticky fiber" sign with erosions[59].

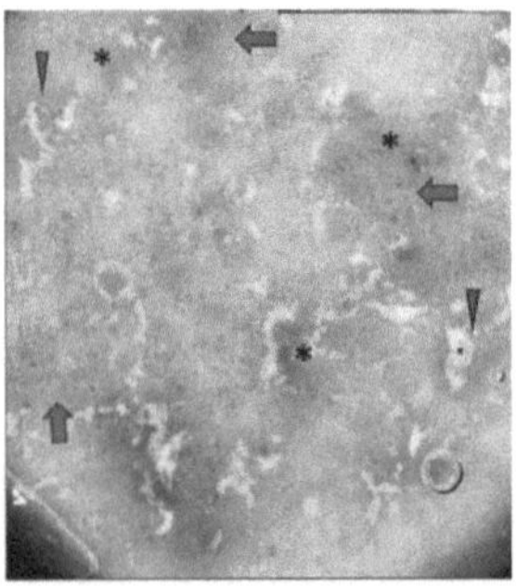

Discoid eczema: dotted vessels in irregular distributions along white scales over yellowish scales[59].

Seborrheic dermatitis:

It's a common encountered relapsing skin condition in the clinical setting that is characterised by scales in the oily based area of the face and anterior chest.
The trichoscope highlight some features, namely; atypical red vessels, twisted red loops, glomerular vessels, thin arborizing red vessels lines, glomerular vessels, structureless red areas, adherent yellowish scales, yellow dots between the follicular opening, perifollicular scales, and dandelion sign which denotes yellow dot surrounded by glomerular and comma vessels which gave the shape described. Also, cherry blossom sign which means arborising vessels with glomerular and comma vessels around them[5-6].

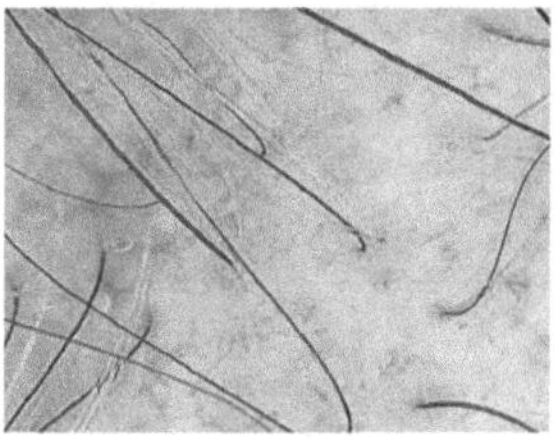

Seborrheic dermatitis: yellow scales [5-6].

Stasis dermatitis:

Exhibits glomerular vessels distributed in clusters or throughout the lesion[41]. Also, displays superficial scales, white shiny structures, multiple rosette signs, white circles and generalized and homogeneous dotted vessels.

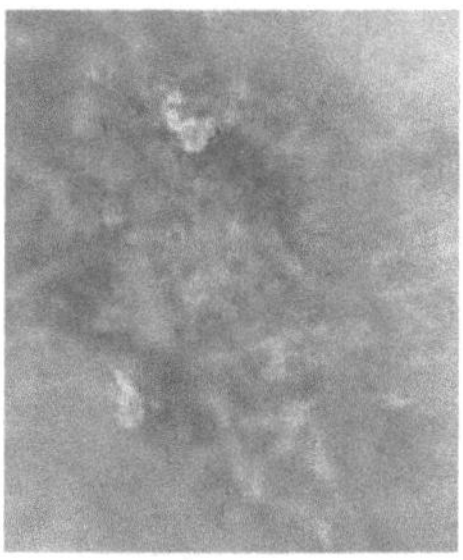

Stasis dermatitis: shows glomerular vessels distributed in clusters or throughout the lesion[41]

Contact dermatitis:

It will show twisted red loops along arborising red lines[5-6].

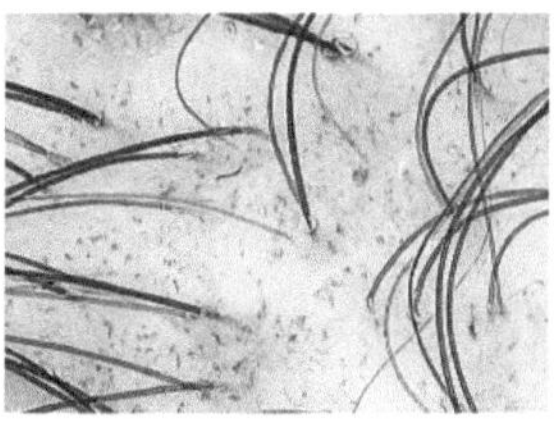

Contact dermatitis: irregular distribution [5-6].

Neurodermatitis/ Lichen simplex chronicus:

It can show erythema, perifollicular scaling associated with hair shaft breakage in a chronic state. Also, on the scalp, an important dermoscopic feature of the disease is the 'broom fibers' which denote a well-delimited hair rarefaction area that displays a tonsure pattern (multiple hair emerging from the same follicle break at the same level distally) [43,57].

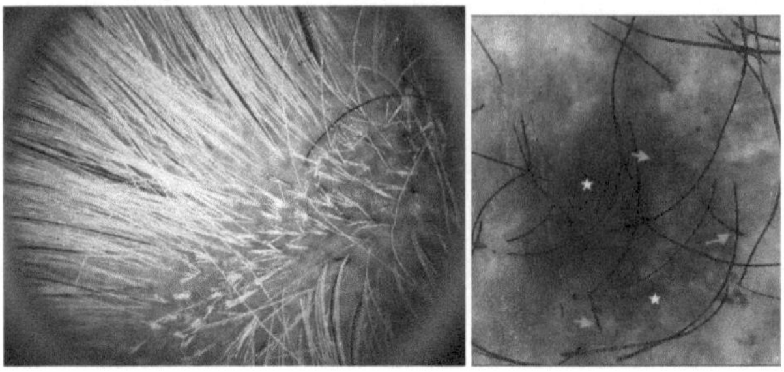

Lichen simplex chronicus: Shows erythema, perifollicular desquamation, and tonsure pattern[43,57].

Acne vulgaris:

The main elementary findings are comedos that appear as dilated follicles filled with a white-yellow circle or a brown plus that coincides with oxidised keratin, papules that appear as reddish, rounded lesions due to the local inflammation, and pustules due to collected purulent material inside the lesion cavity [64].

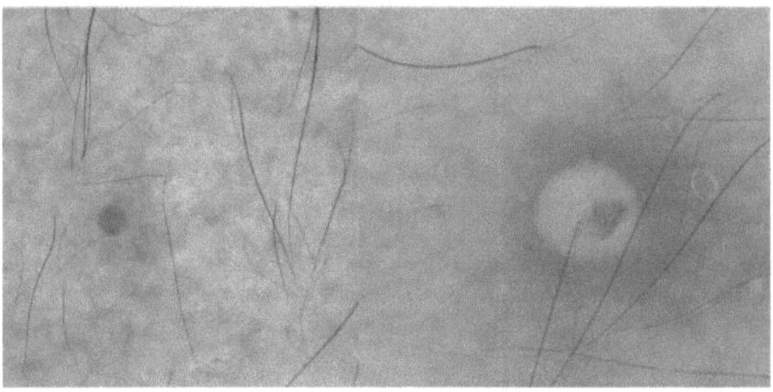

Acne vulgaris: papule and pustule consecutively [64].

Rosacea:

Dermatoscopic sensitive features comprise linear polygonal/ arborising vessels (vascular polygons), orange-yellowish areas along follicular plugs, dilated follicles, and pigmented areas of white scales, which are features of the presence of demodex tails[11-12].

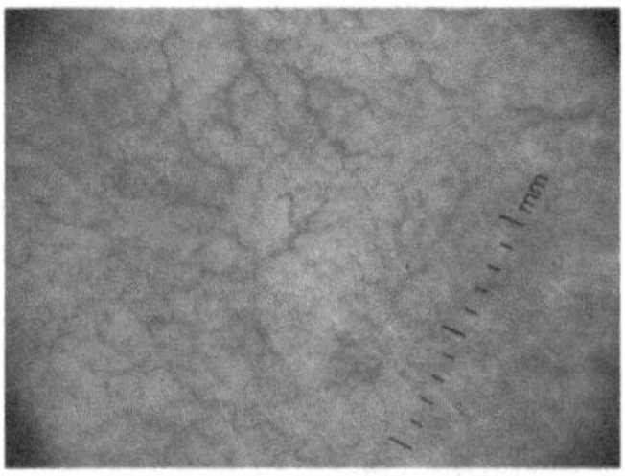

Rosacea: polygon vessels [11-12].

Demodicosis:

Its clinical findings can mimic several dermatoses like rosacea and seborrheic dermatitis. Its dermatoscopic distinctive findings is demodex tails which is white yellow protruding follicular keratotic plugs due to the presence of keratotic material and mites in the follicles [5-7,11].

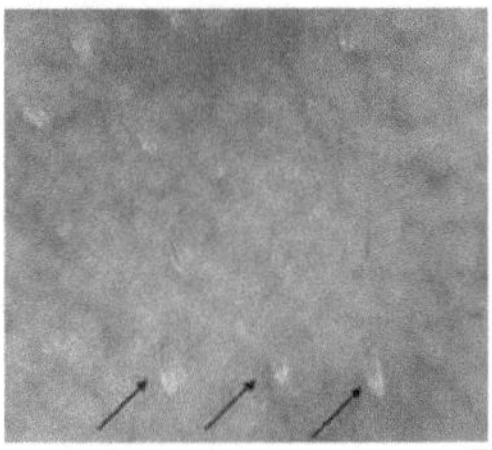

Demodicosis: Demodex tail [7,11].

Discoid lupus erythematosus:

The early stage will show follicular plugs, linear vessels, erythema around follicles, perifollicular white halo, white patches/ scales, honeycomb or brownish pigmentation, structureless brownish areas, red, yellow, and white dots, arborising red lines, loss of follicular units, loss of follicular ostia and perifollicular scaling and pigmentation along telangiectasis at later stages [5-7].

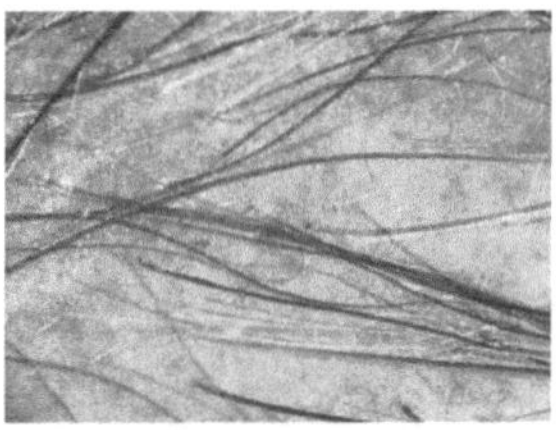

Discoid lupus erythematosus: red brown skin discolouration [5,7].

Subacute lupus erythematosus:

It shows white scales and mixed vascular patterns of linear, linear irregular, branching, and sparsely distributed dotted vessels along pinkish-red background due to hemosiderin deposits, and brown to blue-grey dots/peppering [5-7,13].

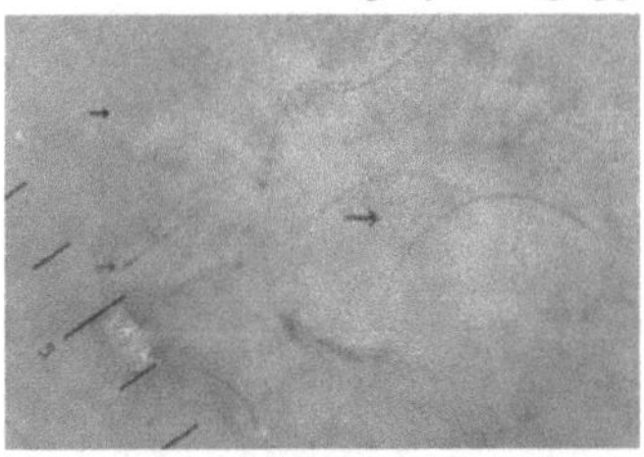

Subacute lupus erythematosus: red-white background, scales, and mixed vascular pattern [5-7].

Lichen planopilaris:

It will show milky red areas or fibrotic patches, perifollicular erythema, pigmented halo in targetoid pattern, loss of vellus hair, perifollicular vessels, black dots, pigmented network, tubular casts, loss of follicular units, white dots, and perifollicular scales [5-7].

Lichen planopilaris: pigmented halo [5-7].

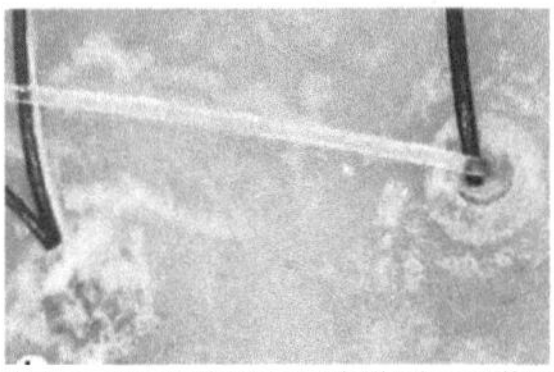

Lichen planopilaris: perifollicular scaling with scales entangling hair shafts [5-7].

Lichen planus:

The main distinctive hallmark is white crossing streaks (Wickham striae) which coincide histologically with hypergranulosis [14]. However, in scaring and resolving dermatosis lesions, a pseudo-Wickham stria can be mimicked in discoid lupus erythematosus, nodular scabies, and prurigo nodularis due to dermal fibrosis, along dotted/ linear vessels, but can be quite dilated than those seen in Wickham striae. However, findings can vary according to lichen planus's variants [5-7].

Also, perifollicular erythema and white halo, interfollicular area blue-gray dots, white dots, and reduced follicular ostia can be seen in follicular lichen planus.

Lichen planus: Wickham striae[14]

Lichen nitidus:

Dermatoscope shows a distinctive round, well-defined white area with an indistinct brown shadow was reflected through each of these white circles (corresponds to epidermal acanthosis), devoid of skin markings due to flattening of the epidermis overlying the inflammatory infiltrate, but on the penis, it retains the skin marking in a pronounced manner [8,51-52].

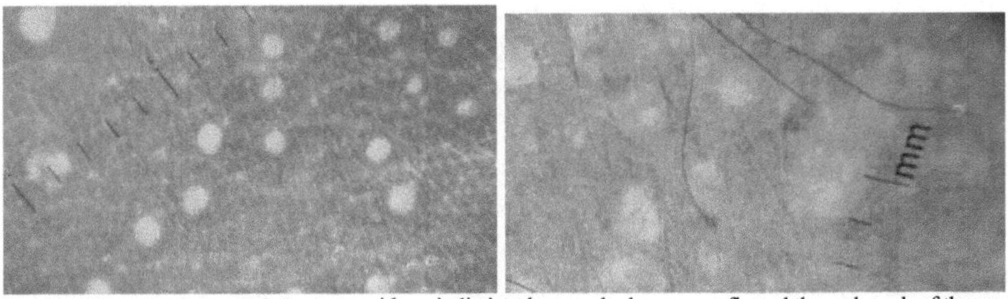

Lichen nitidus: rounded, well-defined area with an indistinct brown shadow was reflected through each of these white circles [14,51,52]

Lichen striatus:

Displays brownish to grayish granular pigmentation, dotted vessels, and white scales[57].

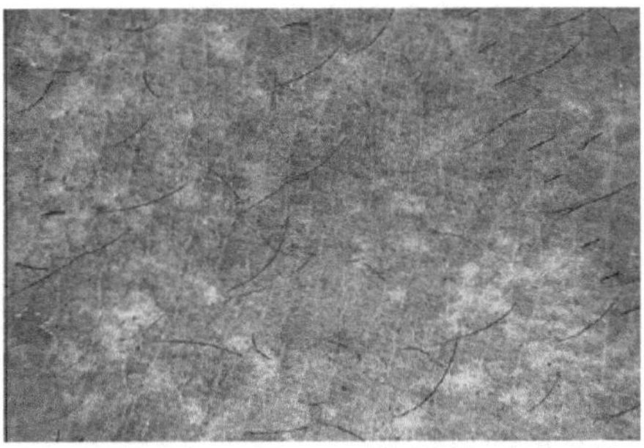
Lichen striatus: brownish to grayish granular pigmentation [57].

Lichen sclerosus et atrophicus (LSA):

keratotic follicular plugs and white structureless areas, corresponding to follicular hyperkeratosis and superficial fibrosis, however, dotted vessels, haemorrhagic dots, erythematous areas, and pigmented dots and networks can be seen in dark skin[57].

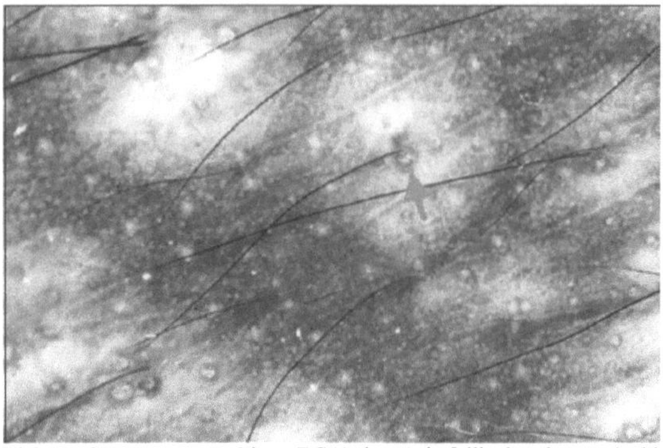
Lichen sclerosus et atrophicus (LSA): keratotic follicular plugs and white structureless areas [57].

Zoon's balanitis (plasma cell balanitis):

Shows homogenous orange structureless areas, corresponding to hemosiderin deposition, and the curved or serpiginous vessels, representing vascular proliferation[57].

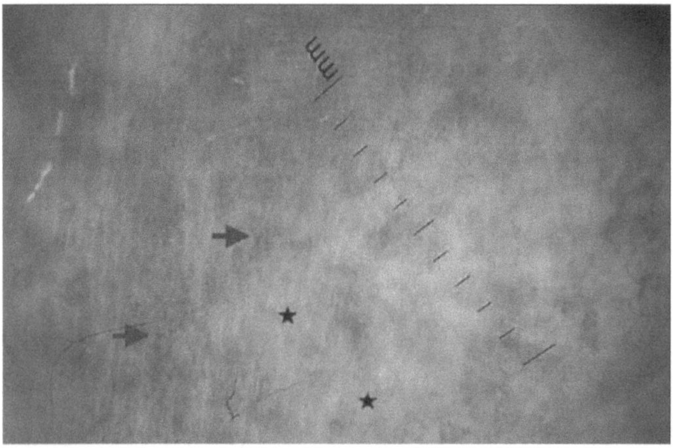
Zoon's balanitis (plasma cell balanitis): homogenous orange structureless areas and curved vessels [57].

Pityriasis rosea:

It displays a yellowish-red background with peripheral white scales (herald collarette sign), along dotted vessels in an irregular patchy manner [7,8].

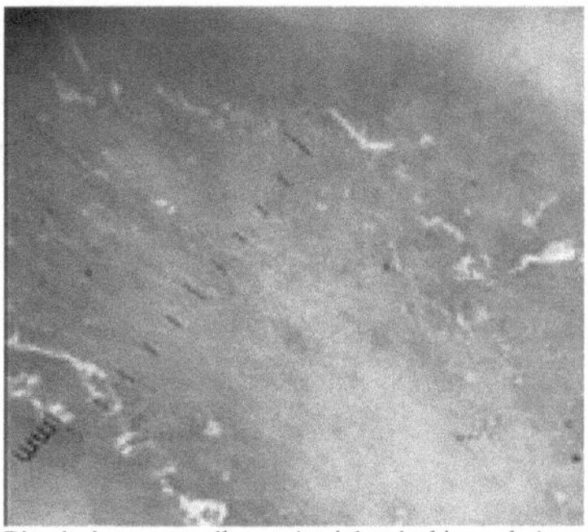

Pityriasis rosea: collarette (peripheral white scales), and dotted vessels[15].

Pityriasis lichenoides:

This encompasses two diseases' spectrum; pityriasis lichenoides et varfioliformis acuta (PLEVA) and pityriasis lichenoides chronica. The latter display orange yellowish structureless areas due to dermal erythrocyte extravasation and non-dotted vessels along the hypopigmented area in long-standing condition, while PLEVA varies according to the lesion age where early one shows purpuric area and a mature lesion show central amorphous brown crust due to epidermal necrosis and in the centre, a white area can be seen in a healing lesion [7,8].

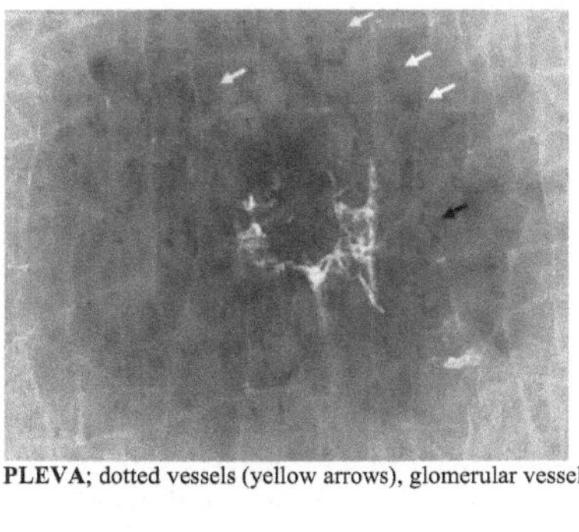

PLEVA; dotted vessels (yellow arrows), glomerular vessels (white arrows), and linear irregular vessels (black arrows) [7-8].

Pityriasis rubra pilaris (PRP):

Yellow area with dotted and linear vessels, with patchy white scales along central follicular keratin plugs, along perifollicular yellow/orange halos, and central hair[18].

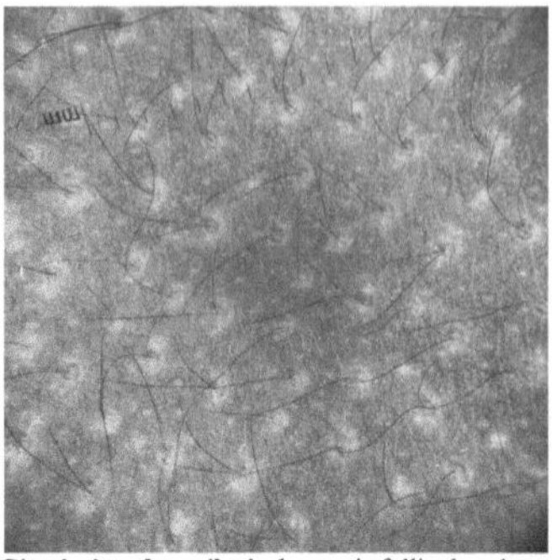

Pityriasis rubra pilaris: keratotic follicular plugs surrounded with erythema and presence of central hair [18].

Porokeratosis:

Coronoid lamella which is the hallmark of porokeratosis histologically, with dermatoscopic finding of well-defined white-yellow annular structure peripherally (white track outlining volcanic crater). In the centre, brown pigmentation, and dotted or linear vessels [17].

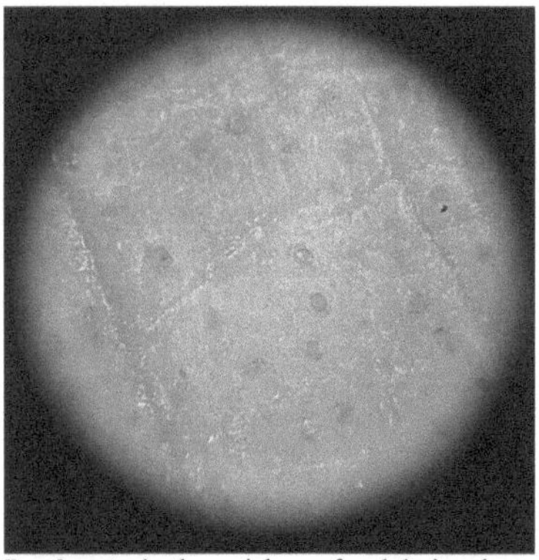

Porokeratosis: the periphery of each lesion shows a white track structure [17].

Darier disease and Grover disease:

Both have similar histology, and thus, they share similar dermatoscope findings.
Both show a central yellow-brown area with star-like branched polygonal or round oval shape due to hyperkeratosis and exocytosis (acantholysis), and a peripheral white halo coincides with acanthosis- pseudocomedones, erythema, dotted or linear vessels [8,9,21,57].

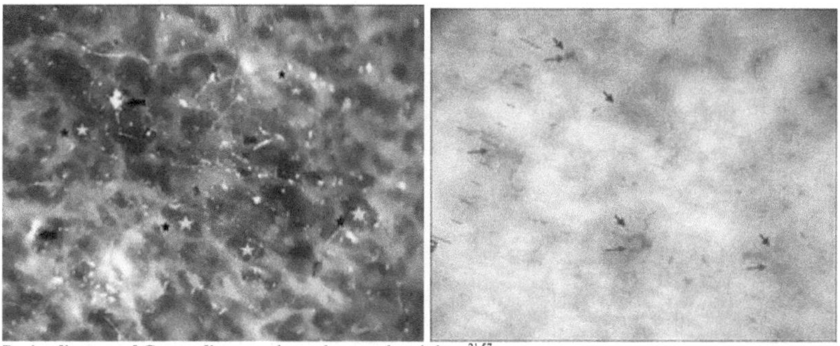

Darier disease and Grover disease: polygonal or round oval shape[21,57]

Hailey- Hailey disease:

Can reveal irregular pinkish-white areas separated by pink furrows, along with whitish areas in a cloud-like arrangement or as irregularly raised wavy folds giving a crumpled fabric pattern. Erosions with a few dotted vessels and crusted areas can be seen[57,58].

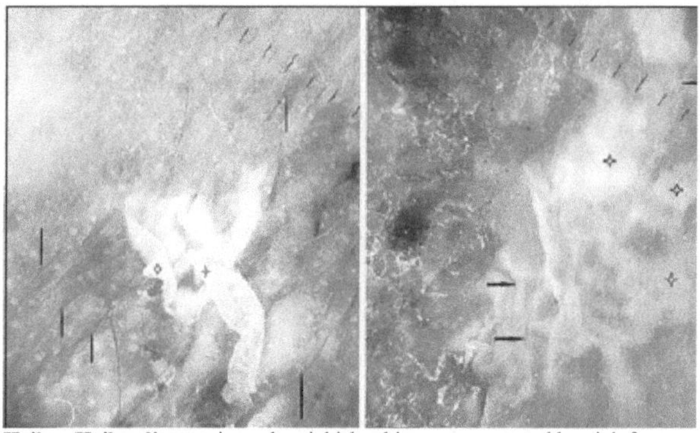

Hailey- Hailey disease: irregular pinkish-white areas separated by pink furrows along with whitish areas [57,58].

Prurigo nodularis:

It shows central erosion, crusting, radially arranged white lines with scales or peripheral white halo with coarse projections on brown-red background (white starburst pattern of pearly white areas with peripheral pigmented dots[57]), haemorrhagic spots, red dots, and globules in excoriated lesions[22].

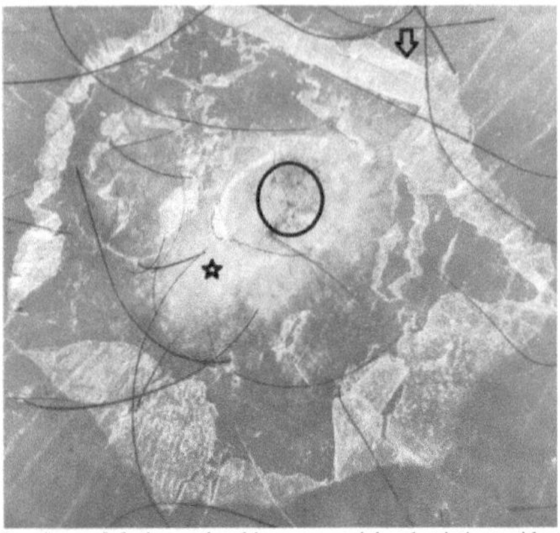

Prurigo nodularis: pearly white areas, peripheral striations with scaling in the hyperkeratotic lesion[22]

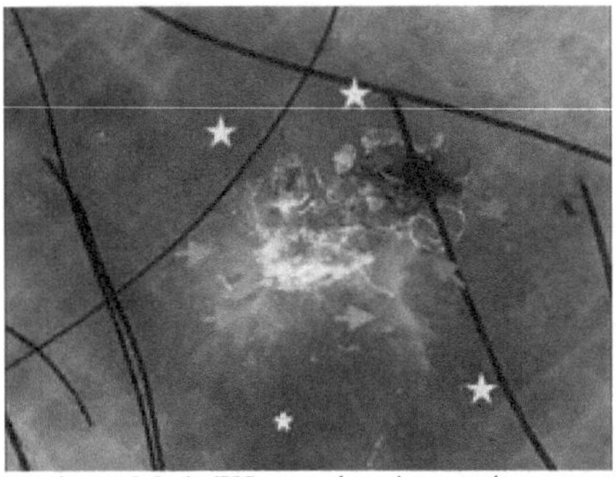

Prurigo nodularis (PN): central erosion, crusting surrounded by a white starburst pattern of pearly white areas[57].

Mastocytosis:

Light brown blot, pigment network, reticular vascular pattern, or yellow orange blot [4-7].

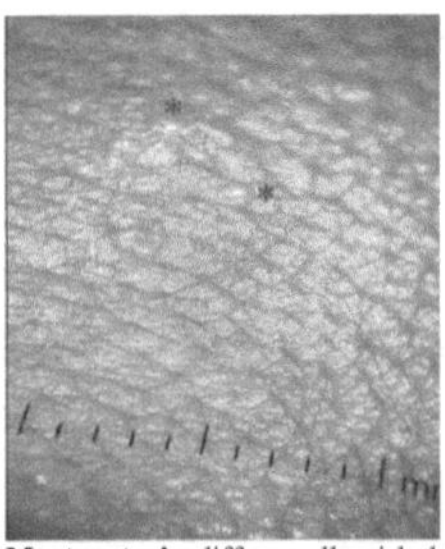

Mastocytosis: diffuse yellowish thickening of the skin giving a "leathery pattern [4-7].

Livedo recticularis:

Linear vessels with a regular distribution, central crusted ulcer, or ivory-white atrophic scar-like areas (corresponding to dermal fibrosis), along the periphery featured hyperpigmentation in a reticular pattern (corresponding to basal layer of the epidermis or melanin within melanophages in dermal papillae), with increased vascular pattern in the periphery including linear telangiectatic vessels and glomerular vessels (correspond to proliferated capillaries in upper dermis)[20].

Livedo recticularis: central crusted ulcer/ white scar-like area with peripheral pigmentation and increased vascularity[20].

Erythema multiform:

The typical target or iris lesions are portrayed by three circular zones: a central dusky zone, surrounded by a ring of pale oedema, and a peripheral red rim. Dermascopy shows Linear vessels peripherally, and bluish patches centrally [25].

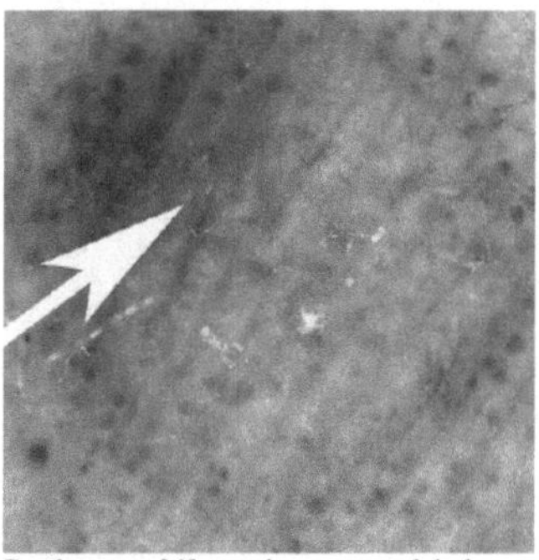

Erythema multiform: shows a central dusky zone of various colours, red/blue/purple and black [25].

Langerhans cell histiocytosis (LCH):

Shows reddish purple areas, brown dots, and a central white area with telangiectasia[57]. The presence of vascular blotch corresponds to dermal haemorrhage, white homogeneous areas to dermal Langerhans cell infiltration without epidermal involvement, and brown dots or structureless area or crust to epidermal infiltration and necrosis by Langerhans cells.

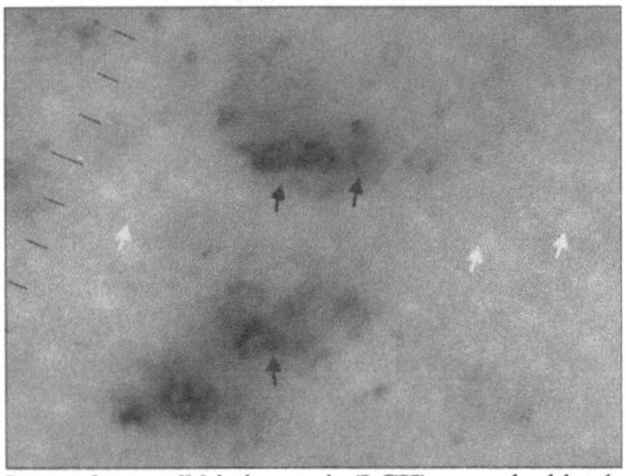

Langerhans cell histiocytosis (LCH): vascular blotches, brown globules, and white structureless area[57].

Sweet syndrome:

Dermoscopy shows structureless bluish patches, with a focal arrangement of red clods vessels [26].

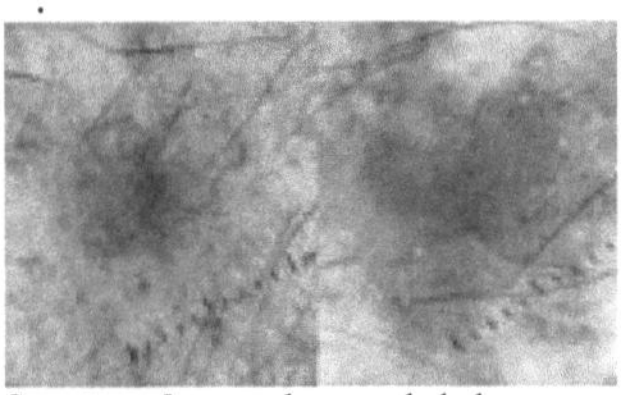

Sweet syndrome: shows red clods over a patchy pinkish background with pale areas[26].

Granuloma;

Dermatoscopic features comprise mainly translucent orange-yellowish globular-like or structureless areas, possibly corresponding to the well-defined sarcoid granulomas predominantly, which imply dermal change with histocyte and lymphocyte density, along linear or branching vessels, and this could fall under sarcoidosis, tuberculosis, lupus vulgaris, and lymphoma [4-7].

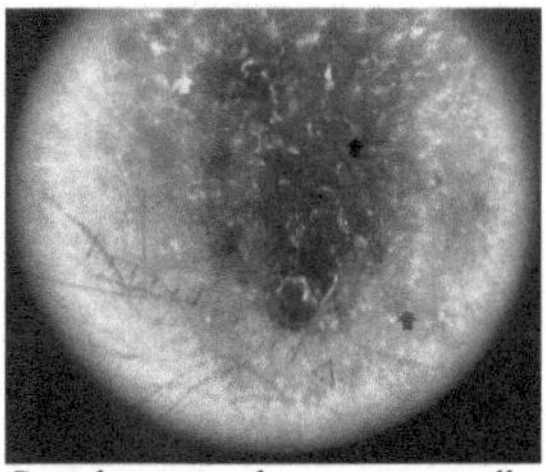
Granuloma: translucent orange-yellowish structuress area [4-7].

Sarcoidosis:

shows orange-yellow globules or areas along linear vessels[27].

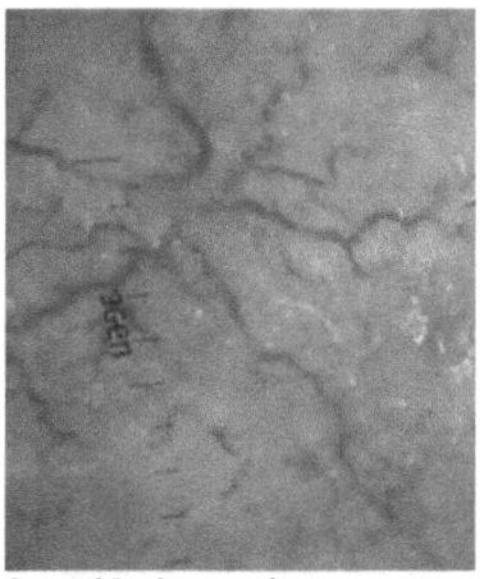
Sarcoidosis: translucent orange yellowish globular-like or structureless areas with linear or branching vessels[27].

Granuloma faciale:

It's not granuloma as the name implies, however it's chronic leukocytoclastic vasculitis, characteristic by the presence of dilated follicular openings, perifollicular white halo, follicular keratotic plugs, linear branching vessels with purpuric spots due to erythrocyte extravasation and hemosiderin dermal deposits[19].

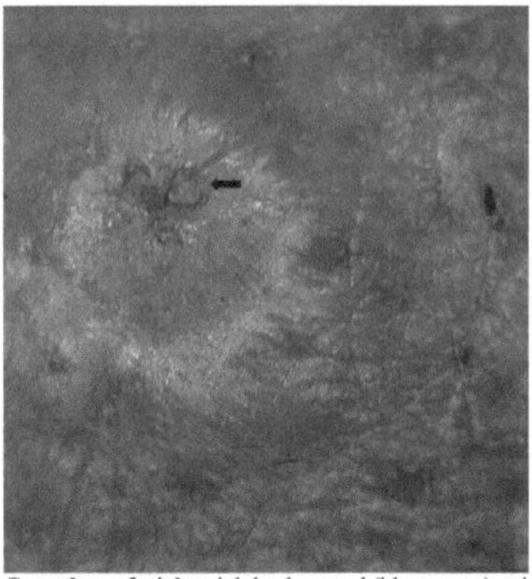

Granuloma faciale: pink background (blue arrow), with aggregates of brown dots/globules (green arrow), white striations in different directions (red arrow), and prominent follicular orifices (black arrow)[19]

Cutaneous leishmaniasis:

An erythema, yellow tear (follicular plugs), hyperkeratosis with central erosion/ ulceration/ crust, with vascular morphology of linear irregular, dotted, hairpin, arborizing, and comma-shaped patterns are mostly appreciated peripherally [38].

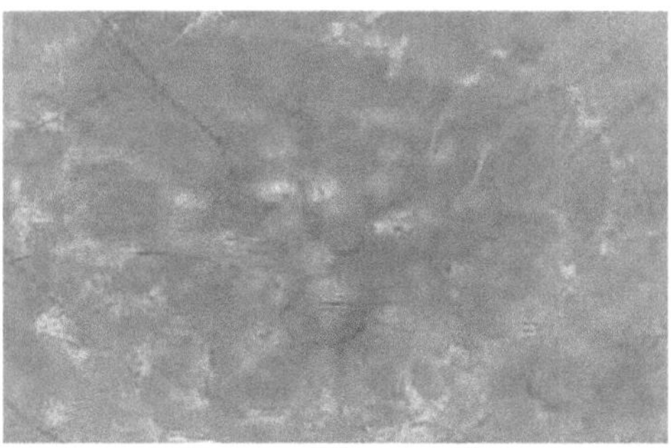

Cutaneous leishmaniasis: showing yellow tears with central crust [38]

Erythema nodosum:

Shows erythematous lobules, shiny white scaling, red dots, hypopigmented structureless areas and vascular dilatation. These lobules substantiated the presence of lobular panniculitides[46-54].

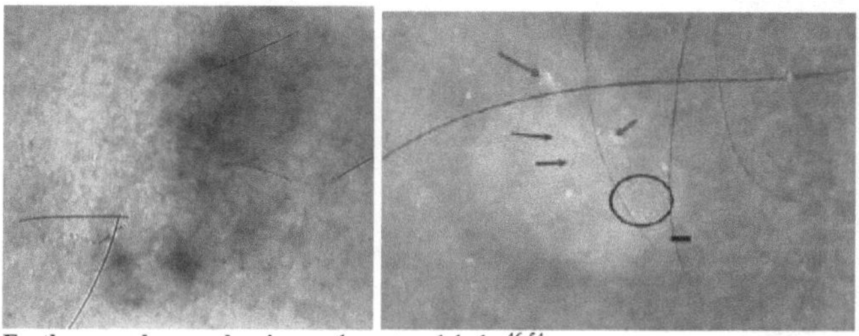

Erythema nodosum: showing erythematous lobules[46-54].

Leprosy:

Tuberculoid/ lepromatous leprosy: shows xerosis, white scaling, and absence of pigment as well as white dots as sweat glands[42].

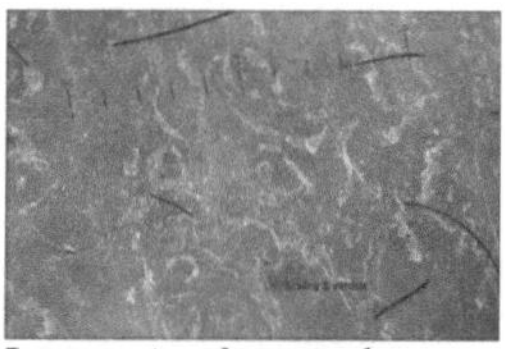

Lepromatous leprosy: shows xerosis and scaling[42].

Tuberculoid leprosy: shows orange-yellowish structureless areas, white structureless areas, peripheral erythema, relative vellus hair sparing, and extensive loss of pigment network54.

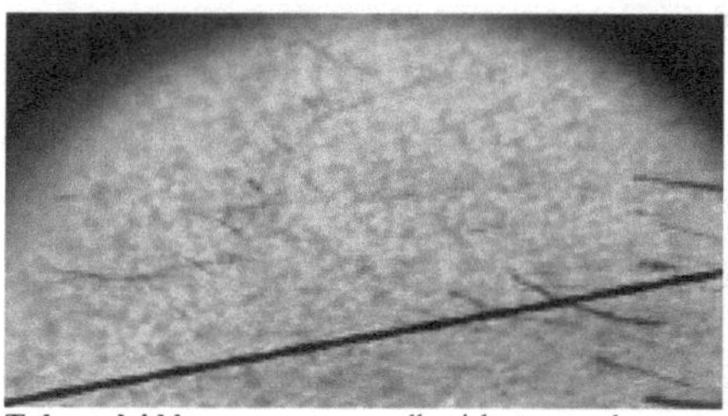

Tuberculoid leprosy: orange-yellowish structureless areas, and extensive loss of pigment network[54].

Borderline Tuberculoid leprosy: shows yellowish orange structureless areas surmounted by branching vessels with violaceous hue in the background, with patchy loss of the pigment network, diminished hair follicles, and sweat glands, along with yellow dots and globules[54].

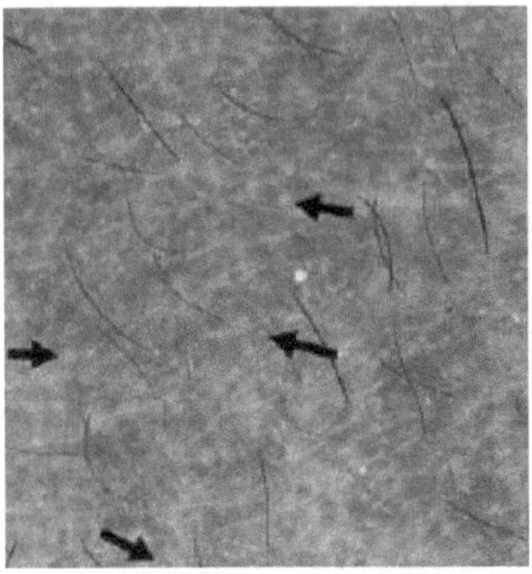

Borderline Tuberculoid leprosy: shows yellowish orange structureless areas with violaceous hue in the background[54].

Lepromatous leprosy:

Yellowish orange and structureless areas with yellow globules, shiny skin, telangiectatic vessels, and sparse appendages along attenuation of the normal reticular pigment network[54].

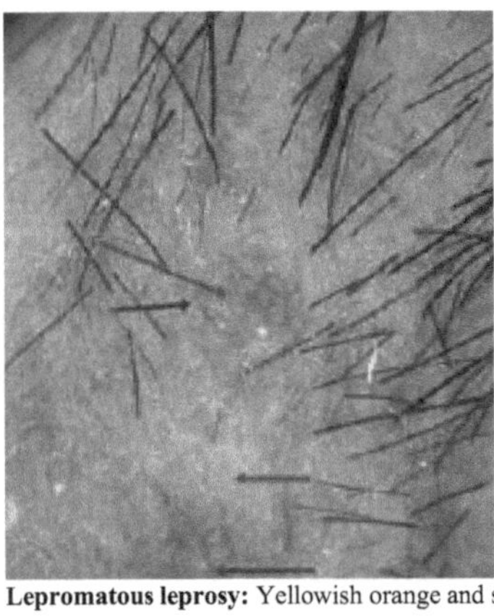

Lepromatous leprosy: Yellowish orange and structureless areas with yellow globules[54].

Borderline Lepromatous Leprosy:

Shows loss of pigment networks with focal areas of hyperpigmentation. Other dermoscopic features included white, shiny streaks with relative sparing of appendages and hair follicles with no branching vessel[54].

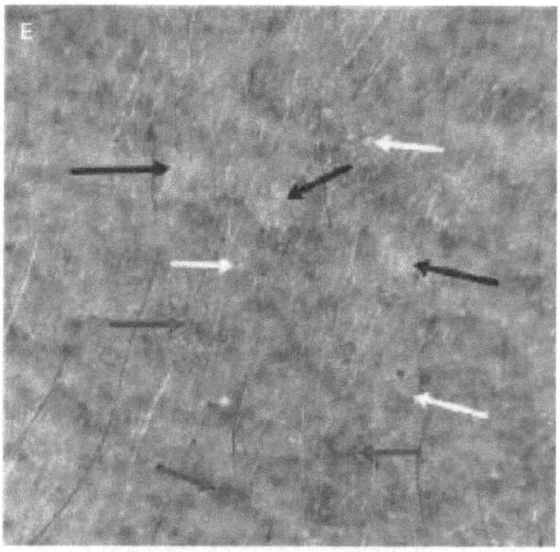

Borderline Lepromatous leprosy: loss of pigment networks with focal areas of hyperpigmentation with relative sparing of appendages and hair follicles[54].

Histoid leprosy:

Exhibited a distinctive dermoscopic appearance, which encompassed crown vessels with central hypopigmented and blanchable dome-shaped structures along with crystalline lines. Additionally, central umbilication displayed central white dots and keratotic plugs along with pseudokoebnerization[54].

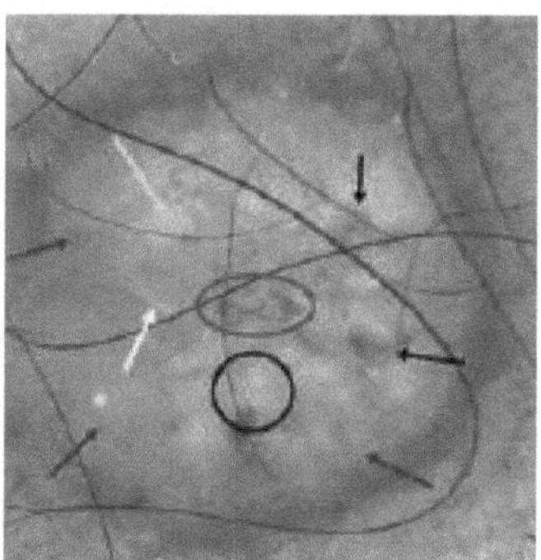

Histoid leprosy: crown vessels, central hypopigmented and blanchable dome-shaped structures, along with crystalline lines and central umbilication of keratotic plug[54].

Granuloma annulare:

Hazy vessels having variable appearance (dotted, linear, short-irregular, and branching) with a background of either red, yellow, or white area, and other findings could be rosettes, crystalline structures, and whitish scaling[34]

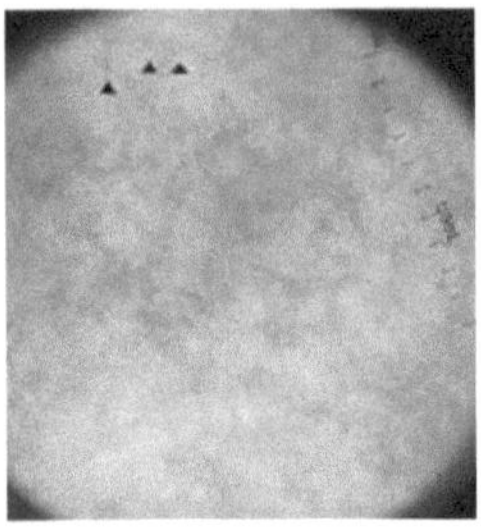

Granuloma annulare: red, yellow or white area [34]

Cutaneous lichen sclerosis and morphoea:

Mostly characterized by bright white yellowish patches/ structureless areas with linear vessels with yellowish keratotic plugs surrounded by a red halo, which marks the disease activity correspondent of dermatosis. In morphoea, pigmentary structures were mostly significant along white clouds, which is an ill-defined, dull white area due to deep dermal fibrosis (white fibrotic beams), linear vessels within a lilac ring are a typical finding[36].

Active cutaneous lichen sclerosis: shows linear branching vessels throughout [36].

Inactive cutaneous lichen sclerosis: shows reticulated brown areas [36].

Morphea: multiple white clouds with pigmentary structure [37].

Necrobiosis lipoidica:

It shows yellowish orange structureless areas due to granuloma inflammatory infiltrate and lipid deposit in the dermis with well-focused vascular structure, along comma-shaped vessels, dotted vessels, globular, comma-shaped and glomerular vessels in earlier lesions stage and hairpin, linear and network like at mature stages or active lesions, or branching serpentine like in advanced stages with attenuation at periphery due to epidermal atrophy of the centre[32].

Necrobiosis lipoidica: irregular arborizing vessels on a light brown background, whitish structures, patchy pigmented reticulum[32].

Urticaria:

Dermoscopy will show enhanced visualization of subclinical red/ purpuric patches of a network of linear vessels surrounding avascular erythematous areas that correspond to dermal oedema[39].

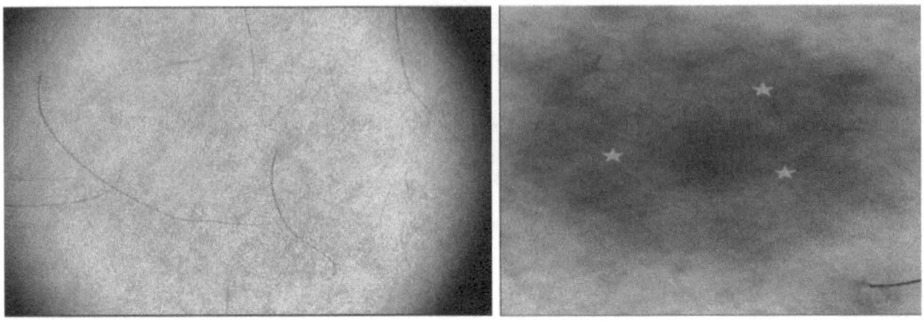

Urticaria: shows a red, reticular network of linear vessels, surrounded by an area devoid of vessels [39].

Urticarial vasculitis:

Purpuric dots or globules, orange-brown background, a network of linear vessels[39].

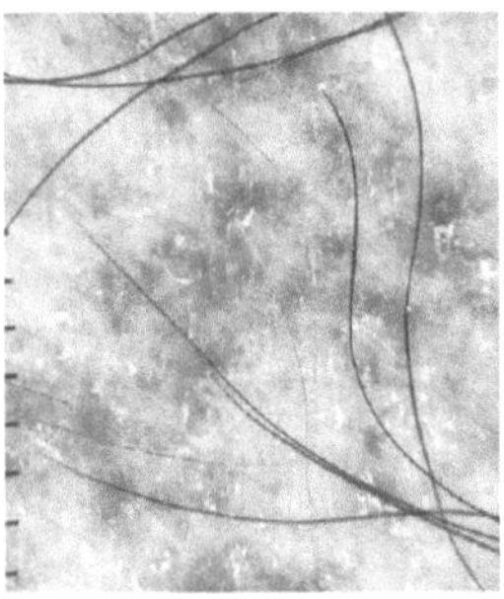

Urticarial vasculitis: shows purpuric globules[39].

Mycosis fungoid, CTCL:

Classic MF shows characteristic by flat macular itchy skin lesions that show dermatoscopically broken small short vessels, dot vessels with yellow crust scales, orange-yellow areas, and spermatozoa-like structures with a striking pigmentary change, thick black lines, white rosettes, and geometric white lines[35].

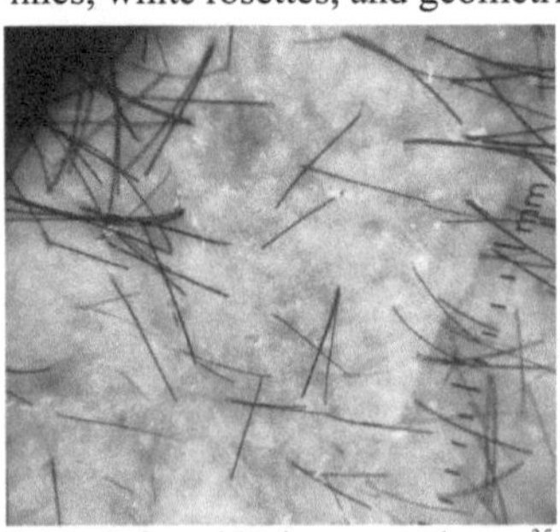

Mycosis fungoid: pigmentary change [35]

Pemphigus foliaceous:

It will show white polygonal structures along linear serpentine vessels, dotted and arborising vessels, linear helical vessels, glomerular vessels, scaling, fried egg sign, yellow haemorrhagic crusts, yellow scaling, tubular casts, haemorrhage, and white perifollicular scaling [4-7].

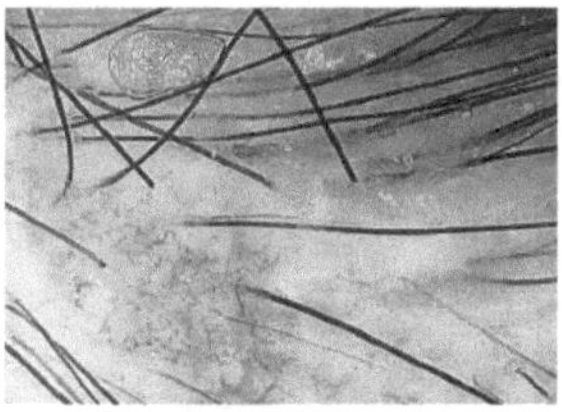

Pemphigus foliaceus: helical and linear serpentine vessels [4-7].

Pemphigus vulgaris:

It will show red dots with white halo and lace-like vessels, a fried egg sign, linear serpentine, lace-like glomerular and arborising vessels, a yellow scale, haemorrhage, dots, and globules [33].

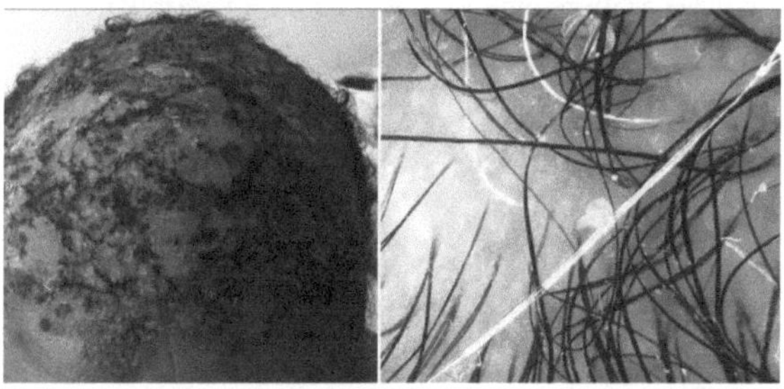

Pemphigus vulgaris: perilesional area reveals hair casts encircling hair shafts [33]

Systemic sclerosis:

It's an autoimmune connective tissue disease that is characterised by: polymorphic vessels, telangiectasia, along salt and pepper signs[28]. Loss of follicular openings, broken hairs, black dots, and pili torti.

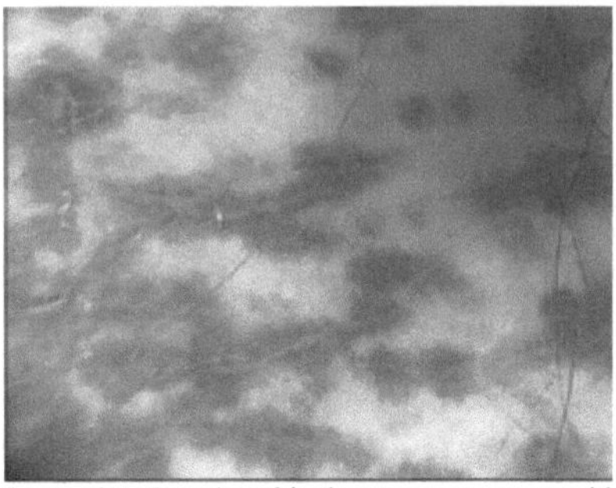

Systemic sclerosis: white homogenous areas with perifollicular pigment,Salt and pepper pigmentation clinically manifests as vitiligo-like depigmentation with perifollicular pigmentary retention[28].

Dermatomyositis:

It's a chronic skin condition with multisystem involvement and some distinctive skin features. Dermsocopy detected advanced nail fold giant capillaries, micro haemorrhages, and avascular areas[29]. The trichoscopic features are: erythematous violaceous areas with enlarged tortuous capillaries or lake-like vascular structures, serpentine vessels, interfollicular scales, peripilar casts, tubular casts, and perifollicular and interfollicular pigmentation, along with scalp atrophy, and tufting of hair where three or more emerging together with peripilar casts[9].

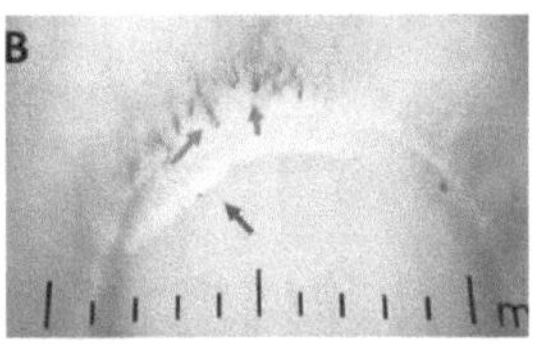

Dermatomyositis: giant nail capillaries [30].

Syphilis:

Dilated capillary vessels, reduced number of terminal hair shafts, broken hairs, thin white scales, perifollicular hyperkeratosis, haemorrhage, yellow dots, pigmented network, white scaling, loss of follicular units, and perifollicular scaling [24].

Syphilis: scaling with skin furrows [24].

Melasma:

It shows a pseudoreticular pigment network, diffuse light-to-dark brown background with sparing of the periappendageal region (follicular and sweat gland openings), brown granules, and globules, including arcuate and annular structures, sometimes with increased vascularization and telangiectasia[31].

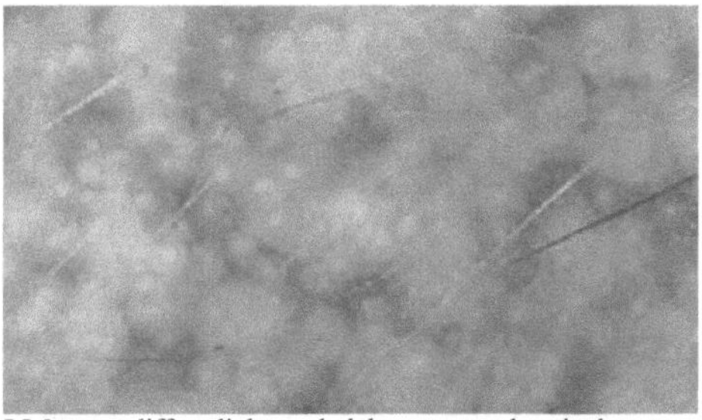

Melasma: diffuse light-to-dark brown pseudoreticular network, multiple brown dots, globules, globule and increased vascularises (red line)[31].

Argyria:

Shows an area comprising blue-grey structures uniformly distributed across a yellow background[53].

Argyria: blue-grey structures uniformly distributed across a yellow background[53].

Pigmented purpuric dermatoses (PPDs):

are a group of inflammatory skin diseases that may be mistaken as vasculitides. dermoscopic features show a copper-red background, rounded to oval dots, grey dots, and a network of brownish-to-grey linked lines, purpuric dots, and an orange-brown area of pigmentation[44].

Pigmented purpuric dermatoses (PPDs): shows discrete yellow red patches with superimposed petechiae[44].

Favre-Racouchot syndrome (FRD):

Showed yellowish lobular-like structures with rare peripheral telangiectasia[55]. In rare for it to display an erythematous background, scar-like depigmentation areas, chrysalides, and fine linear irregular vessels. In the central area, small islands of normal skin could be visualized, while at the periphery of the plaques, there were milia cysts and comedones on[56].

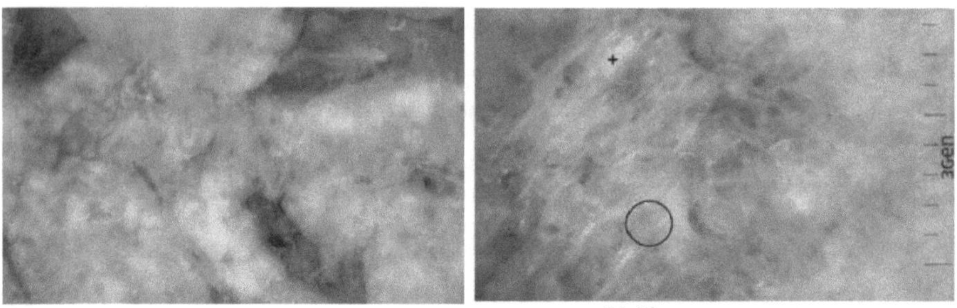

Favre-Racouchot syndrome (FRD): yellowish lobular-like structures[55-56].

Capillaritis:

Is a variant of PPDs and is characterized by extravasation of erythrocytes in the skin (petechia) with hemosiderin deposition over a brownish coppery background, like in the subgroups of PPDs; Schaumberg disease (progressive pigmentary dermatosis), Doucas-Kapetanakis disease, Majocchi disease, Gougerot–Blum syndrome, and lichen aureus[45].

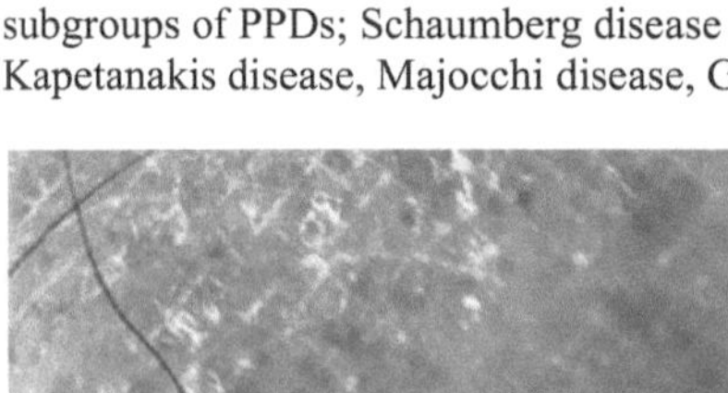

Capillaritis: shows red globules, brown dots and scaling with coppery red background[45].

Cutaneous vasculitis; Three patterns were observed: homogeneous pattern, mottled pattern, and meshy pattern[23].

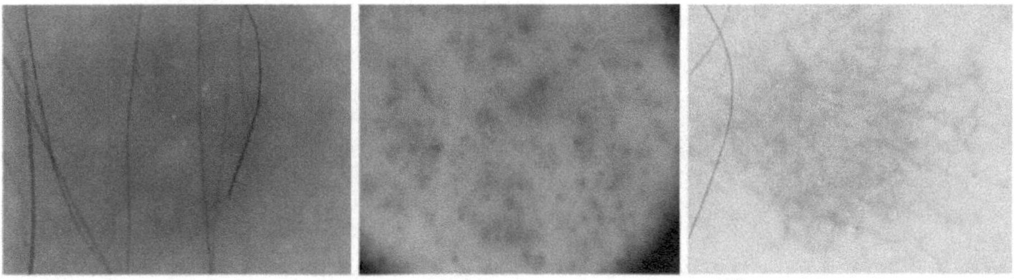

Cutaneous vasculitis: homogeneous pattern, mottled pattern, and meshy pattern [23].

Henoch Schonlein purpura (IgA vasculitis):
Irregularly shaped red patches with blurred borders[23].

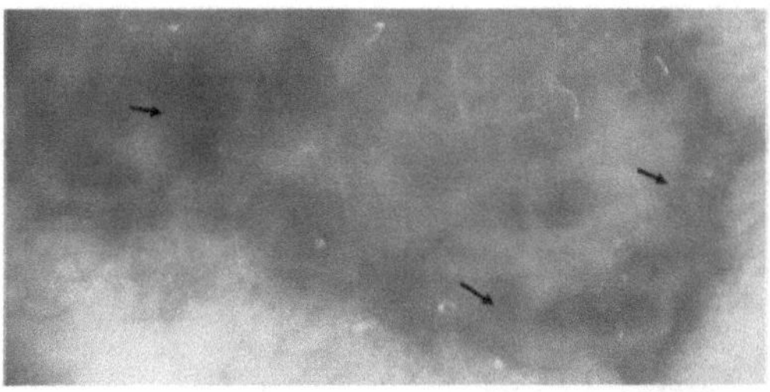

Henoch Schonlein purpura: shows red patches with blurred borders[23].

Diabetic dermopathy:

Central or ring-like globular area surrounded by an ill-defined brown peripheral rim along scarcely branching linear vessels that can look like dusky Wickham striae[40].

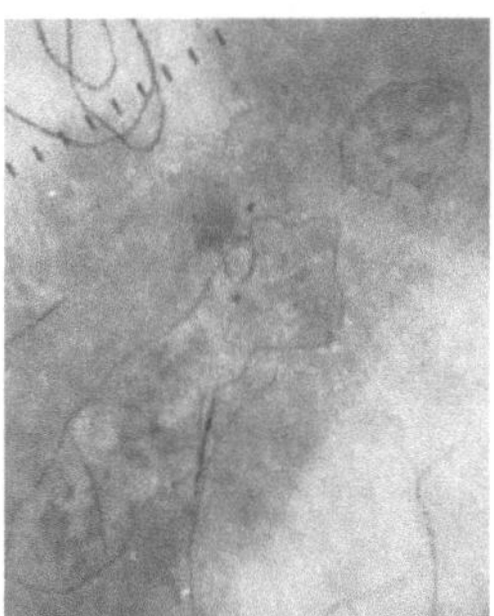

Diabetic dermopathy: ring-like globular structure surrounded by an ill-defined brown peripheral rim[40].

Warts:

Different patterns of finger-like projections with bleeding spots, papillomatous lesions with bleeding spots, and hairpin-like vessels, surrounded by a frogspawn appearance [47,51].

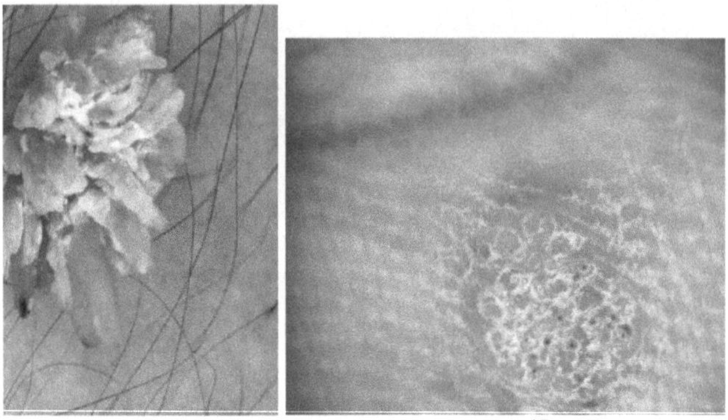

Warts: finger-like projections and frogspawn appearance [47].

Cutaneous Amyloidosis: it can vary and show a central white core with peripheral rippled hyperpigmentation, brown dots, and spoke wheel pigmentation[48].

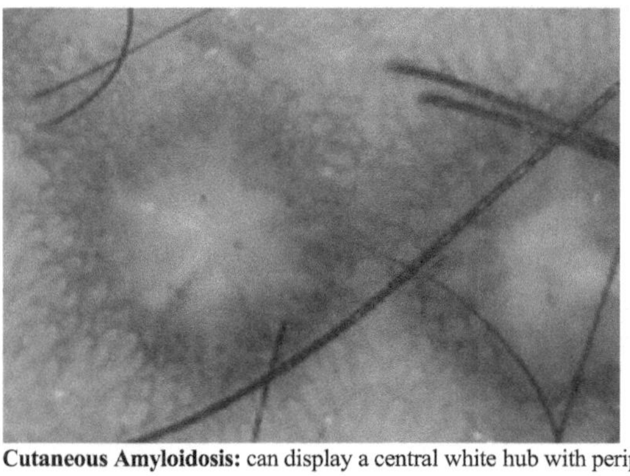

Cutaneous Amyloidosis: can display a central white hub with peripheral hyperpigmentation, and spoke wheel pigmentation[48].

Reactive perforating collagenosis:

Shows a central round yellowish-brown structureless area, corresponding to transepidermal elimination of collagen, surrounded by a white keratotic collarette and erythematous halo- "trizonal concentric" pattern[57].

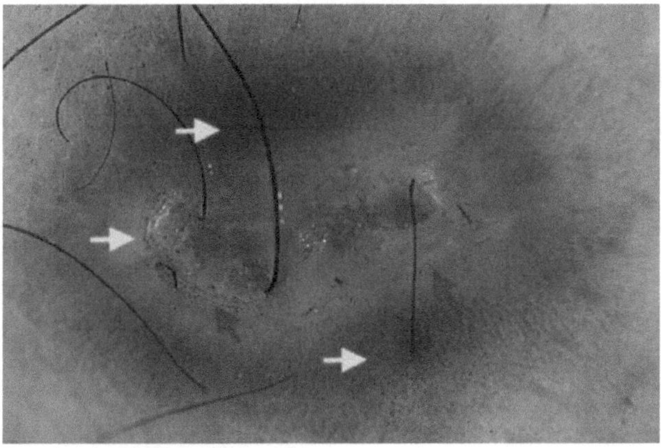

Reactive perforating collagenosis:
central round yellowish-brown structureless area, surrounded by white keratotic collarette and erythematous halo[57].

Keratosis Pilaris:

Can show perifollicular papular erythema, vellus hairs frequently coiled and emerging in groups of 2–3, focal peripilar casts/ scaling, keratotic plugs, white scales, and scattered pigmented globules [49-51].

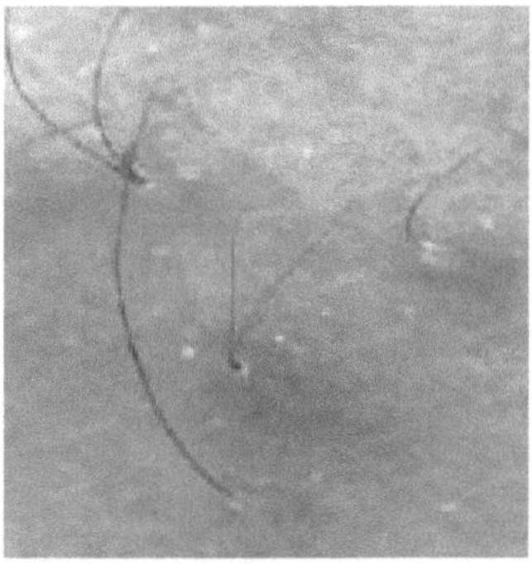

Keratosis Pilaris: shows perifollicular scaling and erythema[50-51].

lichen spinulosus:

perifollicular white halo areas with normal interfollicular areas, perifollicular scaling, and follicular plug [51-52].

lichen spinulosus: perifollicular white halo areas with normal interfollicular areas, and follicular plug [51-52].

Eruptive xanthoma:

Dermoscopy with polarised mode shows a structureless plaque without erythema or vascular proliferation [60].

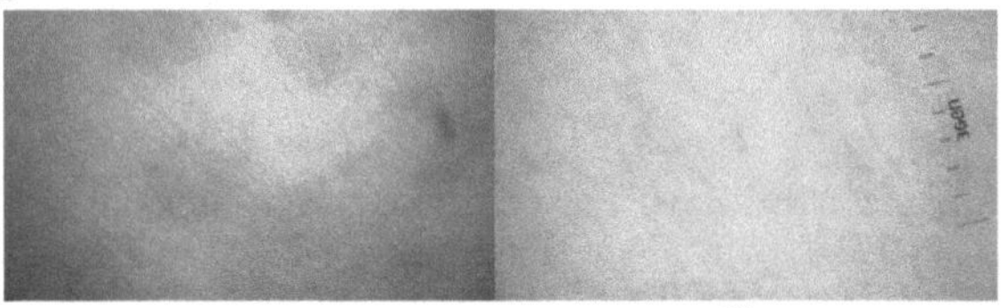

Eruptive xanthoma: shows a structureless plaque [60].

Acanthosis nigricans (AN):

Milder cases show irregular brown globules with perifollicular pigmentation and mild hyperkeratosis and hypermelanization of the basal layer with minimal acanthosis and papillomatosis on histopathology. While facial AN on Dermoscopy shows linear crista cutis, sulcus cutis, and hyperpigmented dots in crista cutis [61].

Acanthosis nigricans: shows hyperpigmented dots in crista cutis [61].

Keloid:

Dermoscopy provides a horizontal view of the upper dermis. The vascular structures of keloids appear linear and relatively parallel to the skin surface, whereas vascular structures are usually not apparent in hypertrophic scars, which helps distinguish keloids and hypertrophic scars even without biopsy [62].

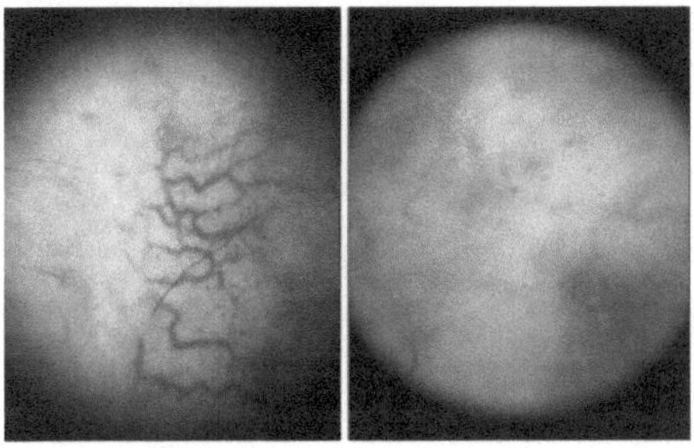

Keloid: distinguish keloids and hypertrophic scars by the vascular structure [62].

Hidradenitis Suppurativa (HS):

Dermoscopy shows multiple keratin accumulations, with irregular openings, either superficially or deeply seated, embedded in whitish cicatricial tissue, and can be observed connected by a blue tunnel [63].

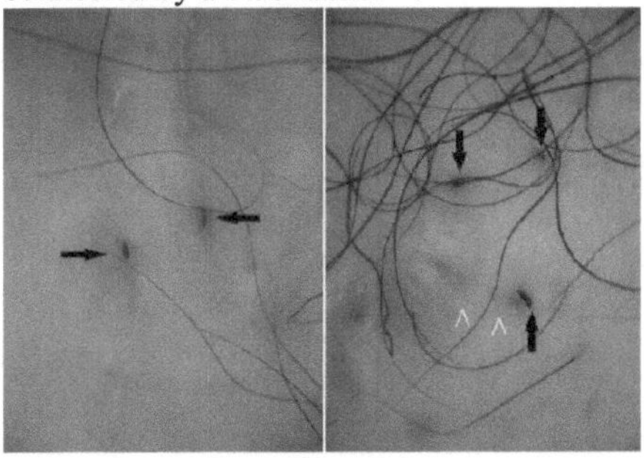

Hidradenitis Suppurativa (HS): multiple keratin accumulation connected by a blue tunnel [63].

Discussion;

Trichoscopy can offer a non-invasive, easier tool in the office for easier diagnosis. It is an important tool in a dermatologist's armamentarium as it can reduce the need for a biopsy in a wide array of conditions[17]. In psoriasis and seborrheic dermatitis, the distinctive scales colour is straightforward, whereas yellowish dominates in seborrheic dermatitis, and in psoriasis, it is whitish [9].
Whereas, in discoid lupus erythematosus and lichen planpilaris, the presence of keratin plugs is the hallmark and specific for discoid lupus, mostly [10].

Conclusion:

Trichoscopy offers a high quality with huge accuracy in the differential diagnosis of inflammatory skin conditions in the clinical setting.

References:

1. Golińska J, Sar-Pomian M, Rudnicka L. Diagnostic Accuracy of Trichoscopy in Inflammatory Scalp Diseases: A Systematic Review. Dermatology. 2022;238(3):412-421. doi: 10.1159/000517516. Epub 2021 Jul 15. PMID: 34265772.
2. Widaty S, Pusponegoro EH, Rahmayunita G, Astriningrum R, Akhmad AM, Oktarina C, Miranda E, Agustin T. Applicability of Trichoscopy in Scalp Seborrheic Dermatitis. Int J Trichology. 2019 Mar-Apr;11(2):43-48. doi: 10.4103/ijt.ijt_86_18. PMID: 31007472; PMCID: PMC6463453.
3. Anna Waśkiel-Burnat, Joanna Czuwara, Leszek Blicharz, Małgorzata Olszewska, Lidia Rudnicka, Differential diagnosis of red scalp: the importance of trichoscopy, *Clinical and Experimental Dermatology*, 2023; llad366, https://doi.org/10.1093/ced/llad366
4. Errichetti E, Ankad BS, Sonthalia S, Jha AK, Keshavamurthy V, Kyrgidis A, Neema S, Chatterjee M, Kaliyadan F, Dogra S, Khare S, Kelati A, Akay BN, Cabo H, Bhat YJ, Bosseila M, Gupta A, Nair P, Gaikwad S, Jayasree P, Sabban ENC, Stinco G, Apalla Z, Zalaudek I, Lallas A. Dermoscopy in general dermatology (non-neoplastic dermatoses) of skin of colour: a comparative retrospective study by the International Dermoscopy Society. Eur J Dermatol. 2020 Dec 1;30(6):688-698. doi: 10.1684/ejd.2020.3928. PMID: 33319764.
5. Errichetti E. Dermoscopy in general dermatology (non-neoplastic dermatoses): pitfalls and tips. Int J Dermatol. 2021 Jun;60(6):653-660. doi: 10.1111/ijd.15427. Epub 2021 Feb 3. PMID: 33533022.
6. Errichetti E, Stinco G. Dermoscopy in General Dermatology: A Practical Overview. Dermatol Ther (Heidelb). 2016 Dec;6(4):471-507. doi: 10.1007/s13555-016-0141-6. Epub 2016 Sep 9. PMID: 27613297; PMCID: PMC5120630.
7. Errichetti E. Dermoscopy of Inflammatory Dermatoses (Inflammoscopy): An Up-to-Date Overview. Dermatol Pract Concept. 2019 Jul 31;9(3):169-180. doi: 10.5826/dpc.0903a01. PMID: 31384489; PMCID: PMC6659598.
8. Shreya K Gowda' Enzo Errichetti, Vishal Thakur, Maitreyee Panda, Siddhartha Dash, Akash Agarwal, Madhusmita Sethy, Pavithra Ayyanar, Biswanath Behera. trichoscopic Features of Scalp Discoid Lupus Erythematosus versus Lichen Planopilaris: A Systematic Review. Clinical, Cosmetic and Investigational Dermatology, 2024,
9. Pozharashka J, Miteva L, Dourmishev L. Cutaneous Manifestations and Their Corresponding Dermoscopic Features in Patients with Dermatomyositis. Dermatol Pract Concept. 2022 Jul 1;12(3):e2022142. doi: 10.5826/dpc.1203a142. PMID: 36159137; PMCID: PMC9464533.
10. Lallas A, Zalaudek I, Argenziano G, Longo C, Moscarella E, Di Lernia V, et al. Dermoscopy in general dermatology. Dermatologic clinics. 2013, 31(4):679-94
11. Kara YA, Özden HK. Dermoscopic Findings of Rosacea and Demodicosis. Indian J Dermatol. 2021 Mar-Apr;66(2):165-168. doi: 10.4103/ijd.IJD_290_18. PMID: 34188272; PMCID: PMC8208267.
12. Stefanou E, Gkentsidi T, et al. Dermoscopic spectrum of rosacea, JEADV. 2022, vol 1 (1) pg 38-44.
13. Behera B, Kumari R, Gochhait D, Ayyanar P. Dermoscopic Appearance of an Annular Subacute Cutaneous Lupus Erythematosus. Dermatol Pract Concept. 2021 Mar 8;11(2): e2021013. doi: 10.5826/dpc.1102a13. PMID: 33747626; PMCID: PMC7943201.

14. Errichetti E. Dermoscopy of inflammatory dermatoses (inflammoscopy): an up-to-date overview. Dermatol Pract Concept.2019;9(3):169-180. DOI: https://doi.org/10.5826/dpc.0903a01
(21) (PDF) Dermoscopy of Inflammatory Dermatoses (Inflammoscopy): An Up-to-Date Overview. Available from: https://www.researchgate.net/publication/335023106_Dermoscopy_of_Inflammatory_Dermatoses_Inflammoscopy_An_Up-to-Date_Overview [accessed Jul 12 2024].
15. Elmas, Ömer Faruk; Kilitçi, Asuman; Acar, Emine Müge. Dermoscopic Profile of Pityriasis Rosea. Dermatologica Sinica 37(4): p 199-204, Oct–Dec 2019. | DOI: 10.4103/ds.ds_14_19
16. Arias-Rodriguez C, Hoyos-Gaviria J, et al. Dermoscopy of pityriasis lichenoides et varioliformis acuta (PLEVA). Clinics Elsevier. 2023, https://clinics.elsevier.es/pt-dermoscopy-pityriasis-lichenoides-et-varioliformis-articulo-S0365059623001964
17. Udare S, Hemmady K. Clinical and dermatoscopic features of porokeratosis palmaris et plantaris. Indian Dermatol Online J. 2016 Jul-Aug;7(4):290-2. doi: 10.4103/2229-5178.185483. PMID: 27559506; PMCID: PMC4976410.
18. Jha AK, Lallas A, Sonthalia S, Jhakar D, Udayan UK, Chaudhary RKP. Differentiation of pityriasis rubra pilaris from plaque psoriasis by dermoscopy. Dermatol Pract Concept. 2018 Oct 31;8(4):299-302. doi: 10.5826/dpc.0804a10. PMID: 30479860; PMCID: PMC6246058.
19. Tandel, Jinal Jainendra; Nair, Pragya Ashok. Granuloma Faciale from Dermatoscopic Perspective. Clinical Dermatology Review 7(1):p 103-106, Jan–Mar 2023. | DOI: 10.4103/cdr.cdr_57_21
20. Hu SC, Chen GS, Lin CL, Cheng YC, Lin YS. Dermoscopic features of livedoid vasculopathy. Medicine (Baltimore). 2017 Mar;96(11):e6284. doi: 10.1097/MD.0000000000006284. PMID: 28296736; PMCID: PMC5369891.
21. Adya KA, Inamadar AC, Palit A. Dermoscopy of Localized Darier's Disease in Fitzpatrick Type IV Skin. Indian Dermatol Online J. 2020 Mar 9;11(2):298-300. doi: 10.4103/idoj.IDOJ_412_18. PMID: 32478011; PMCID: PMC7247640.
22. Supekar, Bhagyashree B.; Wankhade, Vaishali H.; Gawali, Shirin. Rosettes as new dermoscopic finding in prurigo nodularis. Egyptian Journal of Dermatology and Venereology 44(1): p 63-66, January-April 2024. | DOI: 10.4103/ejdv.ejdv_11_23
23. Bakay OSK, Kacar N, Gonulal M, Demirkan NC, Cenk H, Goksin S, Gural Y. Dermoscopic Features of Cutaneous Vasculitis. Dermatol Pract Concept. 2024 Jan 1;14(1): e2024051. doi: 10.5826/dpc.1401a51. PMID: 38364381; PMCID: PMC10868889.
24. Mathur M, Acharya P, Karki A, Shah J, Kc N. Dermoscopic clues in the skin lesions of secondary syphilis. Clin Case Rep. 2019 Jan 25;7(3):431-434. doi: 10.1002/ccr3.1999. PMID: 30899465; PMCID: PMC6406165.
25. Kaliyadan F. Dermoscopy of erythema multiforme. Indian Dermatol Online J. 2017 Jan-Feb;8(1):75. doi: 10.4103/2229-5178.198771. PMID: 28217488; PMCID: PMC5297286.
26. Yumnam D, Melanda H, Batra A, *et al.* unusual presentation of Sweet's syndrome involving palms and its dermoscopy. *BMJ Case Reports CP* 2022;15: e249352.
27. Ramadan S, Hossam D, Saleh MA. Dermoscopy could be useful in differentiating sarcoidosis from necrobiotic granulomas even after treatment with systemic steroids. Dermatol Pract Concept. 2016 Jul 31;6(3):17-22. doi: 10.5826/dpc.0603a05. PMID: 27648379; PMCID: PMC5006548.

28. Bains A, Rajagopal SV, Lahoria U. Dermoscopy of Salt and Pepper Pigmentation. Indian Dermatol Online J. 2020 Sep 19;13(4):548-550. doi: 10.4103/idoj.IDOJ_333_20. PMID: 36262586; PMCID: PMC9574142.
29. Pozharashka J, Miteva L, Dourmishev L. Cutaneous manifestations and their corresponding dermoscopic features in patients with dermatomyositis Dermatol Pract Concept. 2022;12(3): e2022142. DOI: https://doi.org/10.5826/dpc.1203a142
30. Dan J, Sprow G et al. Characterizing nailfold capillary changes in dermatomyositis with a dermatoscope. Arthritis Rheumatol.2022; 74(supp9).
31. Sonthalia S, Jha AK, Langar S. Dermoscopy of Melasma. Indian Dermatol Online J. 2017 Nov-Dec;8(6):525-526. doi: 10.4103/idoj.IDOJ_6_17. PMID: 29204416; PMCID: PMC5707865.
32. Montero C, Izquierdo J et al, 2013, dermoscopy of necrobiosis lipoidica, to the editor, vol 104, no 6, 534-537
33. Pirmez, Rodrigo. Acantholytic Hair Casts: A Dermoscopic Sign of Pemphigus Vulgaris of the Scalp. International Journal of Trichology 4(3): p 172-173, Jul–Sep 2012. | DOI: 10.4103/0974-7753.100087
34. Errichetti E, Lallas A, Apalla Z, Di Stefani A, Stinco G. Dermoscopy of Granuloma Annulare: A Clinical and Histological Correlation Study. Dermatology. 2017;233(1):74-79. doi: 10.1159/000454857. Epub 2017 Jan 19. PMID: 28099955.
35. Nakamura M, Huerta T, Williams K, Hristov AC, Tejasvi T. Dermoscopic features of mycosis fungoides and its variants in patients with skin of color: a retrospective analysis. Dermatol Pract Concept. 2021;11(3): e2021048. DOI: https://doi.org/10.5826/dpc.1103a48
36. Errichetti E, Lallas A, Apalla Z, Di Stefani A, Stinco G. Dermoscopy of Morphea and Cutaneous Lichen Sclerosus: Clinicopathological Correlation Study and Comparative Analysis. Dermatology. 2017;233(6):462-470. doi: 10.1159/000484947. Epub 2017 Dec 14. PMID: 29237158.
37. Wang YK, Hao JC, Liu J, Liu YH, Jin HZ. Dermoscopic features of morphea and extragenital lichen sclerosus in Chinese patients. Chin Med J (Engl). 2020 Sep 5;133(17):2109-2111. doi: 10.1097/CM9.0000000000000977. PMID: 32769495; PMCID: PMC7478797.
38. Serarslan G, Ekiz Ö, Özer C, Sarıkaya G. Dermoscopy in the Diagnosis of Cutaneous Leishmaniasis. Dermatol Pract Concept. 2019 Apr 30;9(2):111-118. doi: 10.5826/dpc.0902a06. PMID: 31106013; PMCID: PMC6502294.
39. Madarkar, Mahajabeen Saheb Patel. Importance of Dermoscopy in Differentiating Common Urticaria and Urticarial Vasculitis: A Randomized Control Study. Clinical Dermatology Review 5(2): p 144-148, Jul–Dec 2021. | DOI: 10.4103/CDR.CDR_82_20
40. Navarro-Gilo M, Alberti-Masalleras N, de Gea-Fernández M, Domingo A, Sagristà M. Dermoscopy of Diabetic Dermopathy. Dermatol Pract Concept. 2022 Oct 1;12(4):e2022185. doi: 10.5826/dpc.1204a185. PMID: 36534534; PMCID: PMC9681172.
41. Zaballos P, Salsench E, Puig S, Malvehy J. Dermoscopy of venous stasis dermatitis. Arch Dermatol. 2006 Nov;142(11):1526. doi: 10.1001/archderm.142.11.1526. PMID: 17116857.
42. Debdeep Mitra, Ajay Chopra, 1427. Correlation of Dermoscopic and Histopathologic Patterns in Leprosy-a Pilot Study, *Open Forum Infectious Diseases*, Volume 9, Issue Supplement_2, December 2022, ofac492.1256, https://doi.org/10.1093/ofid/ofac492.1256

43. Muylaert BPB, Borges MT, Michalany AO, Scuotto CRC. Lichen simplex chronicus on the scalp: exuberant clinical, dermoscopic, and histopathological findings. An Bras Dermatol. 2018 Jan-Feb;93(1):108-110. doi: 10.1590/abd1806-4841.20186493. PMID: 29641708; PMCID: PMC5871373.
44. Spigariolo CB, Giacalone S, Nazzaro G. Pigmented Purpuric Dermatoses: A Complete Narrative Review. J Clin Med. 2021 May 25;10(11):2283. doi: 10.3390/jcm10112283. PMID: 34070260; PMCID: PMC8197337.
45. Singh J, Sahu P, Dayal S, Kataria SP. Unilateral Linear Capillaritis Dermoscopic Examination: A Distinct Clinico-Histopathological Correlation. Dermatol Pract Concept. 2021 Jul 8;11(3): e2021053. doi: 10.5826/dpc.1103a53. PMID: 34414008; PMCID: PMC8337041.
46. Chuh A et al. the first application of epiluminescence dermoscopy in erythema nodosum. Our Dermatol online. 2018; 9(3): 282-284.
47. Al Rudaisat M, Cheng H. Dermoscopy Features of Cutaneous Warts. Int J Gen Med. 2021 Dec 16; 14:9903-9912. doi: 10.2147/IJGM.S335276. PMID: 34938109; PMCID: PMC8687684.
48. Dincy Peter CV, Agarwala MK, George L, Balakrishnan N, George AA, Mahabal GD. Dermoscopy in Cutaneous Amyloidosis. - A Prospective Study from India. Indian J Dermatol. 2022 Jan-Feb;67(1):94. doi: 10.4103/ijd.ijd_850_20. PMID: 35656239; PMCID: PMC9154142.
49. Sonthalia S, Bhatia J, Thomas M. Dermoscopy of Keratosis Pilaris. Indian Dermatol Online J. 2019 Aug 28;10(5):613-614. doi: 10.4103/idoj.IDOJ_279_18. PMID: 31544095; PMCID: PMC6743401.
50. Gangadhar M, Adya KA, Inamadar AC. A Study of Clinical, Dermoscopic and Histopathological Correlation in Follicular Keratotic Diseases: Preliminary Observations in 30 Cases. Indian Dermatol Online J. 2021 Aug 2;12(5):731-740. doi: 10.4103/idoj.IDOJ_96_21. PMID: 34667761; PMCID: PMC8456266.
51. Kansal NK, Vasisht S and Bhatia R, 2024. Dermoscopic features of lichen nitidus, BMJ case Rep, 17 e259998.doi;10.1136.
52. Malakar S, Save S, Mehta P. Brown Shadow in Lichen Nitidus: A Dermoscopic Marker! Indian Dermatol Online J. 2018 Nov-Dec;9(6):479-480. doi: 10.4103/idoj.IDOJ_338_17. PMID: 30505802; PMCID: PMC6232985.
53. Enei ML, Paschoal FM, Valdés R. Argyria mimicking a blue nevis: dermoscopy features. An Bras Dermatol. 2013 May-Jun;88(3):452-5. doi: 10.1590/abd1806-4841.20132022. PMID: 23793194; PMCID: PMC3754384.
54. Mohta A, Jain SK, Agrawal A, Kushwaha RK, Sharma P, Sethia K, Jain M. Dermoscopy in Leprosy: A Clinical and Histopathological Correlation Study. Dermatol Pract Concept. 2021 Apr 12;11(2): e2021032. doi: 10.5826/dpc.1102a32. PMID: 33954015; PMCID: PMC8060002.
55. Chessa MA, Filippi F, Ferrara F, Patrizi A, Baraldi C. A case of unilateral inflamed plaques with comedones of the face. Dermatol Pract Concept. 2018 Oct 31;8(4):292-294. doi: 10.5826/dpc.0804a07. PMID: 30479857; PMCID: PMC6246051.
56. Melo PV, Schoenardie BO, Escobar GF, Bonamigo RR. Dermoscopic Features of an Actinic Comedonal Plaque - a Rare Ectopic Form of Favre-Racouchot Syndrome. Dermatol Pract Concept. 2023 Jan 1;13(1): e2023034. doi: 10.5826/dpc.1301a34. PMID: 36892333; PMCID: PMC9946079.
57. Bhat, Yasmeen J.; Jha, Abhijeet K.[1]. Dermatoscopy of Inflammatory Diseases in Skin of Color. Indian Dermatology Online Journal 12(1): p 45-57, Jan–Feb 2021. | DOI: 10.4103/idoj.IDOJ_613_20

58. Chauhan P, Meena D, Hazarika N. Dermoscopy of Hailey Hailey Disease. Indian Dermatol Online J. 2018 Mar-Apr;9(2):139-140. doi: 10.4103/idoj.IDOJ_202_17. PMID: 29644210; PMCID: PMC5885629.
59. Żychowska, M.; Kołcz, K. Dermoscopy for the Differentiation of Subacute Cutaneous Lupus Erythematosus from Other Erythematous Desquamative Dermatoses—Psoriasis, Nummular Eczema, Mycosis Fungoides and Pityriasis Rosea. *J. Clin. Med.* 2024, *13*, 577. https://doi.org/10.3390/jcm13020577.
60. Wollina U, Koch A, Hansel G, Schönlebe J. Eruptive Xanthomas - Two Case Reports with Distinct Features. Open Access Maced J Med Sci. 2018 Nov 21;6(11):2152-2154. doi: 10.3889/oamjms.2018.458. PMID: 30559880; PMCID: PMC6290429.
61. Shah VH, Rambhia KD, Mukhi JI, Singh RP, Kaswan P. Clinico-investigative Study of Facial Acanthosis Nigricans. Indian Dermatol Online J. 2022 Mar 3;13(2):221-228. doi: 10.4103/idoj.IDOJ_855_20. PMID: 35287408; PMCID: PMC8917481.
62. Yoo MG, Kim IH. Keloids and hypertrophic scars: characteristic vascular structures visualized by using dermoscopy. Ann Dermatol. 2014 Oct;26(5):603-9. doi: 10.5021/ad.2014.26.5.603. Epub 2014 Sep 26. PMID: 25324653; PMCID: PMC4198588.
63. Francesco LACARRUBBA, Maria Letizia MUSUMECI, Maria Rita NASCA, Anna Elisa VERZÌ, Federica FIORENTINI and Giuseppe MICALI. Double-ended Pseudocomedones in Hidradenitis Suppurativa: Clinical, Dermoscopic, and Histopatho- logical Correlation.2016, December. Short communication, doi: 10.2340/00015555-2601.
64. Alma A, Sticchi A, Chello C, Guida S, Farnetani F, Chester J, Bettoli V, Pellacani G, Manfredini M. Dermoscopy, Reflectance Confocal Microscopy and Optical Coherence Tomography Features of Acne: A Systematic Review. J Clin Med. 2022 Mar 24;11(7):1783. doi: 10.3390/jcm11071783. PMID: 35407391; PMCID: PMC8999263.

TABLE OF CONTENTS

Introduction	3
Discussion	36
Conclusion	36
References	37

Kolibri 360 GmbH
Ferdinand-Jühlke-Straße 7
99095 Erfurt, Deutschland
produktsicherheit@kolibri360.de

Druck:
CPI Druckdienstleistungen GmbH
im Auftrag der
Kolibri 360 GmbH
Ferdinand-Jühlke-Str. 7
99095 Erfurt